MIX
Papier aus verantwortungsvollen Quellen
Paper from responsible sources
FSC® C105338

Harikumar Rajaguru
Sunil Kumar Prabhakar

Comprehensive Analysis of Extreme Learning Machine and Continuous Genetic Algorithm for Robust Classification of Epilepsy from EEG Signals

Anchor Academic Publishing

Rajaguru, Harikumar, Prabhakar, Sunil Kumar: Comprehensive Analysis of Extreme Learning Machine and Continuous Genetic Algorithm for Robust Classification of Epilepsy from EEG Signals, Hamburg, Anchor Academic Publishing 2017

Buch-ISBN: 978-3-96067-099-5
PDF-eBook-ISBN: 978-3-96067-599-0
Druck/Herstellung: Anchor Academic Publishing, Hamburg, 2017

Bibliografische Information der Deutschen Nationalbibliothek:
Die Deutsche Nationalbibliothek verzeichnet diese Publikation in der Deutschen Nationalbibliografie; detaillierte bibliografische Daten sind im Internet über http://dnb.d-nb.de abrufbar.

Bibliographical Information of the German National Library:
The German National Library lists this publication in the German National Bibliography. Detailed bibliographic data can be found at: http://dnb.d-nb.de

All rights reserved. This publication may not be reproduced, stored in a retrieval system or transmitted, in any form or by any means, electronic, mechanical, photocopying, recording or otherwise, without the prior permission of the publishers.

Das Werk einschließlich aller seiner Teile ist urheberrechtlich geschützt. Jede Verwertung außerhalb der Grenzen des Urheberrechtsgesetzes ist ohne Zustimmung des Verlages unzulässig und strafbar. Dies gilt insbesondere für Vervielfältigungen, Übersetzungen, Mikroverfilmungen und die Einspeicherung und Bearbeitung in elektronischen Systemen.

Die Wiedergabe von Gebrauchsnamen, Handelsnamen, Warenbezeichnungen usw. in diesem Werk berechtigt auch ohne besondere Kennzeichnung nicht zu der Annahme, dass solche Namen im Sinne der Warenzeichen- und Markenschutz-Gesetzgebung als frei zu betrachten wären und daher von jedermann benutzt werden dürften.

Die Informationen in diesem Werk wurden mit Sorgfalt erarbeitet. Dennoch können Fehler nicht vollständig ausgeschlossen werden und die Diplomica Verlag GmbH, die Autoren oder Übersetzer übernehmen keine juristische Verantwortung oder irgendeine Haftung für evtl. verbliebene fehlerhafte Angaben und deren Folgen.

Alle Rechte vorbehalten

© Anchor Academic Publishing, Imprint der Diplomica Verlag GmbH
Hermannstal 119k, 22119 Hamburg
http://www.diplomica-verlag.de, Hamburg 2017
Printed in Germany

Abstract

Epilepsy is a common and diverse set of chronic neurological disorders characterized by seizures. It is a paroxysmal behavioral spell generally caused by an excessive disorderly discharge of cortical nerve cells of brain. Epilepsy is marked by the term "epileptic seizures". Epileptic seizures result from abnormal, excessive or hyper synchronous neuronal activity in the brain. About 50 million people worldwide have epilepsy, and nearly 80% of epilepsy occurs in developing countries. The most common way to interfere the epilepsy is to analysis the EEG (electroencephalogram) signal which is non invasive, multi channel recording of the brain's electrical activity. It is also essential to classify the risk levels of the epilepsy so that the diagnosis can be made easy. This project investigates the possibility of Extreme Learning Machine (ELM) and Continuous GA as a post classifier for detecting and classifying the epilepsy of various risk levels from the EEG signals. The Singular Value Decomposition (SVD), Principal Component Analysis (PCA) and Independent Component Analysis (ICA) are used for dimensionality reduction.

Table of Contents

Abstract 1
1. Introduction 5
2. Materials and Methods 6
3. Extreme Learning Machine (ELM) 11
4. Continuous Genetic Algorithm as a Post Classifier for Epilepsy
   Risk Level Classification 16
      4.1 Functions and Operators 17
      4.2 Mean Square Error 19
5. Results and Discussion 21
      5.1 Performance Index 21
      5.2 Quality Value 22
6. Conclusion 29
References 30

1. Introduction

This project is to analysis the performance of the Extreme Learning Machine and Genetic Algorithm for the classification of the epilepsy from the EEG signals. The Singular Value Decomposition, Principal Component Analysis and Independent Component Analysis are used for Dimensionality Reduction. The neural activity of the human brain starts between the 17th and 23rd week of prenatal development. It is believed that from this early stage and throughout life electrical signals generated by the brain represent not only the brain function but also the status of the whole body. Epilepsy is a common and diverse set of chronic neurological disorders characterized by seizures. It is a paroxysmal behavioral spell generally caused by an excessive disorderly discharge of cortical nerve cells of brain and can range from clinically undetectable (electrographic seizures) to convulsions.

Epileptic seizures result from abnormal, excessive or hyper synchronous neuronal activity in the brain. About 50 million people worldwide have epilepsy, and nearly 80% of epilepsy occurs in developing countries. Epilepsy becomes more common as people age. Onset of new cases occurs most frequently in infants and the elderly. Epileptic seizures may occur in recovering patients as a consequence of brain surgery. Epilepsy is usually controlled, but not cured, with medication. However, more than 30% of people with epilepsy do not have seizure control even with the best available medications. Surgery may be considered in difficult cases. Not all epilepsy syndromes are lifelong – some forms are confined to particular stages of childhood. Epilepsy should not be understood as a single disorder, but rather as syndromic with vastly divergent symptoms, all involving episodic abnormal electrical activity in the brain and numerous seizures.

Understanding of neuronal functions and neurophysiological properties of the brain together with the mechanisms underlying the generation of signals and their recordings is, however, vital for those who deal with these signals for detection, diagnosis, and treatment of brain disorders and the related diseases. The genesis of the EEG signal and the epilepsy detection from the EEG signals are discussed as follows.

2. Materials and Methods

The EEG signal acquisition is the first step for the classification of epilepsy. The EEG signal of twenty patients are analysed in this study. A paper record of 16 channel EEG data is obtained from Sri Ramakrishna hospital, Coimbatore, India. The EEG signal was band pass filtered between 0.5 Hz and 50Hz using five pole analog Butter worth filters z to remove the artifacts. With the help of neurologist, the artifacts free EEG signals are selected since the artifacts free signals enable us to have accurate detection of epilepsy.The EEG signal data are obtained for duration of about 30 seconds and they are divided into epochs of 2 seconds. Because the 2 second epochs are long enough to detect any significant changes in the electrical activity of the brain and presence of artifacts and short enough to avoid any repetition or redundancy in the signal [1],[2],[3].

The EEG signal has a maximum frequency of 50Hz and so, each epoch is sampled at a frequency of 200Hz. Each sample corresponds to the instantaneous amplitude values of the signal, totalling 400 values for an epoch. Each channel has 400 samples of EEG signals per epoch and four such epochs of data forms a bin. There are sixteen such a bins are available per patient. The data volume for a patient is around 25,600 samples. Hence this large amount of data necessitates the dimensionality reduction technique level for processing the EEG signal.

DIMENSIONALITY REDUCTION TECHNIQUES

The functional block diagram of the three epilepsy risk level classifier is shown in Fig.2.1. The EEG data used in the study were acquired from twenty epileptic patients who had been under the evaluation and treatment in the Neurology department of Sri Ramakrishna Hospital, Coimbatore, India. A paper record of 16 channel EEG data is acquired from a clinical EEG monitoring system through 10-20 international electrode placing method. With an EEG signal free of artifacts, a reasonably accurate detection of epilepsy is possible; however, difficulties arise with artifacts. This problem increases the number of false detection that commonly plagues all classification systems. With the help of neurologist, we had selected artifact free EEG records with distinct features.

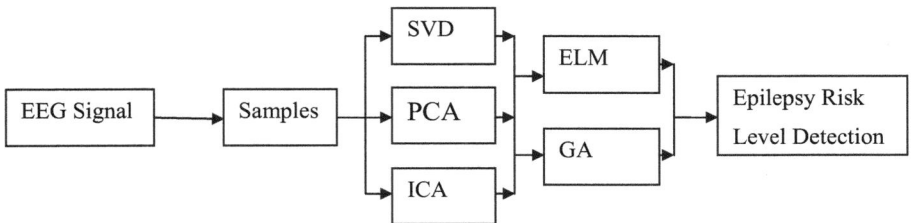

Figure 1: Block Diagram for epilepsy detection

Epileptic EEG is acquired and the dimensionality reduction is performed by Singular Value Decomposition, Principal Component Analysis and Independent Component Analysis. A matrix of (20 x 20) data is reduced to a matrix of (20 x 1) using SVD, PCA and ICA. The final risk level classification is done by Extreme Learning machine and Genetic Algorithm. The twenty patients are labeled into nine groups based upon their EEG signal and clinical observation. The patients are assigned and labeled in groups based on the severity index or target values of the frequent occurrence of epileptic seizures. The patients who are labeled in the G_1 group are considered to be more vulnerable one and G_9 group as least disturbed.

Table 2.1: Target values for Groups

GROUP	PATIENT NUMBER	TARGET VALUE
G_1	1,2,11,20	0.85 (High Risk)
G_2	3,7,14,18	0.65
G_3	4,13	0.45
G_4	6,12	0.35
G_5	5,8,19	0.25
G_6	15,17	0.20
G_7	16	0.15
G_8	10	0.1
G_9	9	0.05 (Low Risk)

1. Singular Value Decomposition (SVD)

The SVD is a factorization of a real or complex matrix with useful applications in signal processing. It is the method for identifying and ordering the dimensions along which the data points exhibit the most variation. This leads to the way that, once the variation in data points is identified, it is possible to find the best approximation of the original data points using fewer dimensions and hence SVD is used to reduce a large matrix (invertible and square matrix) into significantly small matrix[4].

The mathematical definition of SVD is given as follows:

Let X denote a m × n matrix of real value data. The equation for singular value decomposition of X is given as [5]

$$X = USV^T \qquad (1)$$

Where U is the m × n matrix, S is an n × n diagonal matrix and V^T is also an n × n matrix. The columns of U are called the left singular vectors, $\{u_k\}$. The rows of V^T are the right singular vectors. Each s_i is equal to $\sqrt{\lambda_i}$, the square root of eigen values of $C = X^T X$. The elements of S are nonzero on the diagonal, and they are the singular values of the matrix X. That is

$$S = \text{diag}(s_1, \ldots s_n) \qquad (2)$$

It is assumed that $s_1 \geq s_2 \geq \ldots \geq s_n$ for convince. Thus SVD takes a high dimensional, high variable set of data points and reduces it to a lower dimensional space that exposes the substructure of the original data. The reason that makes SVD for NLP applications is that SVD simply ignores the variation below a particular threshold to reduce the data.

2. Principal Component Analysis (PCA)

The PCA is the technique to find the patterns in the high dimensional data. Once the patterns are identified, it is easy to compress the data by reducing the number of dimensions without much information loss [6]. The reduction in dimension has the following advantages;

    i) The computational overhead of the subsequent process is reduced.
    ii) noise may be reduced

This statistical technique has found application in various fields such as data compression, image processing, visualization, exploratory data analysis, pattern recognition and time series prediction [7].

The basic idea of PCA is to find the principal components $s_1, s_2, \ldots s_n$ so that they explain the maximum amount of variance possibly by n linearly transformed components. PCA can be defined using the recursive formulation. The direction of the first principal component w_1 is defined as

$$w_1 = \genfrac{}{}{0pt}{}{\arg\max}{\|w\|=1} E\{(w^T x)^2\} \qquad (3)$$

where w_1 is of the same direction as the random data vector x. Thus the first principal component is the projection on the direction in which the variance of the projection is maximized. Having determined the first k-1 principal components, the k-th principal component is determined as the principal component of the residual:

$$w_k = \genfrac{}{}{0pt}{}{\arg\max}{\|w\|=1} E\left\{\left[w^T\left(x - \sum_{i=1}^{k-1} w_i\, w_i^T x\right)\right]^2\right\} \qquad (4)$$

The principal components can be now calculated as

$$s_i = w_i^T x \qquad (5)$$

The w_i are the eigen vectors corresponding to the n largest eigen values of the covariance matrix C, calculated as

$$E\{xx^T\} = C \qquad (6)$$

3. Independent Component Analysis (ICA)

ICA of a random vector consists of searching for the linear transformation that minimizes the statistical dependence between the components. ICA has applications in the areas of data compression, blind identification & deconvolution and localization of sources. In this model, the data variables are assumed to be linear mixtures of some unknown latent variables and the mixing system is also unknown. These latent variables are assumed to be non- Gaussian and mutually independent called independent components. They are also called as sources or factors. ICA is capable of finding the underlying factors or sources from the data variables even when the classical methods such as PCA and factor analysis fail completely [8], [9].

ICA can be defined using a statistically 'latent variable' model [10]. Assume that we observe n linear mixtures $x_1,...,x_n$ of independent components

$$x_j = a_{j1}s_1 + a_{j1}s_2 + ... + a_{jn}s_n, \text{ for } j=1,n \tag{7}$$

The time index t is not considered here. Because each mixture x_j, and the independent component s_k are random variables. Without loss of generality, both the mixture variables and the independent components are assumed to have zero mean i.e.

$$\hat{x} = x - E(x) \tag{8}$$

The convenient method is to use the vector notation instead of the above equations. Let **x** be the random vector with the elements as mixtures $x_1,...,x_n$ and s be the random vector with the elements $s_1, s_2,...s_n$. Generally the bold lower case letters indicate vectors and bold upper-case letters denote matrices. Let **A** be the matrix containing the elements a_{ij} which can be modelled as

$$\mathbf{x} = \mathbf{As} \tag{9}$$

By denoting the above equation by a_{ij} for the column matrix of **A**, the above equation can also be written as

$$\mathbf{x} = \sum_{i=1}^{n} \mathbf{a}_i s_i \tag{10}$$

The above equation is called the Independent Component Analysis or ICA model.

3. Extreme Learning Machine(ELM)

The traditional feed forward neural network parameters need to be tuned and due to this, dependency between the different layers of parameters exists. It is known that the gradient descent based methods have been used in various algorithms of feed forward neural network such as BPNN. But due to the improper learning steps, the learning methods of gradient descent based methods are very slow or may easily converge to the local minima. Also the neural networks may be over trained by using these algorithms thus obtaining a worst generalization performance [11].

In order to obtain to obtain the high classification accuracy with less training time, Huang *et al* in [11]-[13] ,[17], [18] proposed new learning algorithm called the Extreme Learning Machine for single –hidden layer feed forward neural networks. The extreme learning machine is high popular due to its high generalization ability. The ELM is used to classify the protein sequence classification in [14] with ten class of super families obtained from a domain database. On comparing the result of the ELM with that of Back –Propagation Neural Networks, the ELM outperforms the BPNN.

In [15], R. Zhang et al developed an ELM for multi category classification in three Cancer Microarray Gene Expression datasets and the results prove that ELM can also avoid problems such as over-fitting, local minima, and improper learning rate. Apart from the field of bioinformatics, ELM has been applied to Biosignal Processing. N. Y. Liang , et al in [16] proposed ELM based classification scheme to classify five mental tasks from different subjects using EEG signals. The performance of the ELM with BPNN and Support Vector Machine(SVM) is compared and the results show that ELM needs an orders of magnitude less training time compared to SVMs and two orders of magnitude less training time compared to BPNN. According to G.Geetha et al , the existing learning techniques of SLFN can be applied only to the differentiable activation functions whereas, the ELM algorithm can also be used for non-differentiable activation functions.

The ELM algorithm is explained as follows: Suppose learning N arbitrary different instances (x_i, t_i) where $X_i = [x_{i1}, x_{i2}, \ldots, x_{in}]^T \in R^n$ and $t_i = [t_{i1}, t_{i2}, \ldots, t_{im}]^T \in R^m$, standard Single-layer Feedforward Networks with N hidden neurons and activation function $g(x)$ are mathematically modelled as a linear system as

$$\sum_{i=1}^{\tilde{N}} \beta_i g(w_i \cdot x_j + b_i) = T_j \tag{11}$$

Where $w_i = [w_{i1}, w_{i2}, \ldots, w_{in}]^T$ denotes the weight vector connecting the ith hidden neuron and the input neuron, $\beta_i = [\beta_{i1}, \beta_{i2}, \ldots, \beta_{in}]^T$ denotes the weight vector connecting the i-th hidden neuron and output neurons, and b_i represents the threshold of the i-th hidden neuron. $w_i * w_j$ represents the inner product of w_i and x_j. If the Single-layer Feedforward Network with N hidden neurons with activation function $g(x)$ is able to approximate N distinct instances (x_i, t_i) with zero error means that

$$H\beta = T \tag{12}$$

Where

$H(w_1, \ldots, w_{Nh}, b_1, \ldots, b_{Nh}, x_1, \ldots, x_N) =$

$$\begin{bmatrix} g(w_1 \cdot x_1 + b_1) & \cdots & g(w_{Nh} \cdot x_1 + b_{Nh}) \\ \vdots & \vdots & \vdots \\ g(w_1 \cdot x_N + b_1) & \cdots & g(w_{Nh} \cdot x_N + b_{Nh}) \end{bmatrix} \tag{13}$$

and

$$W = \begin{bmatrix} w_1^T \\ \vdots \\ w_{Nh}^T \end{bmatrix}_{Nh \times m} \quad T = \begin{bmatrix} t_1^T \\ \vdots \\ t_N^T \end{bmatrix}_{N \times m} \tag{14}$$

H is the hidden layer output matrix of the SLFN. Hence for fixed arbitrary input weights w_i and the hidden layer bias s, training a Single-layer Feed-forward Network equals to discovering a least-squares solution $\hat{\beta}$ of the linear system $H\beta = T$, $\hat{\beta} = H^\dagger T$ is the best weights, where $H^\dagger$ is the Moore-Penrose generalized inverse.

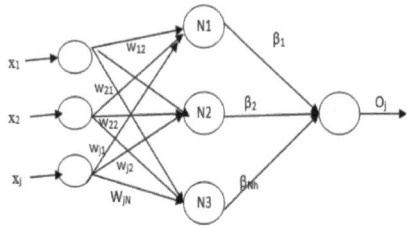

Figure 2: The Structure of ELM

The sigmoid function with a gain parameter λ is used in training instead of threshold function directly as in [18] and it is given by

$$g(x) = 1/(1 + \exp(-\lambda x)) \quad (15)$$

The sine function is given by

$$g(x) = 0.5 \sin(x) \quad (16)$$

The hard limiting function is calculated as

$$\text{hardlim}(n) = 1 \text{ if } n \geq 0$$
$$= 0 \text{ o.w} \quad (17)$$

The triangular basis function is given by

$$\text{tribas}(n) = 1 - \text{abs}(n), \text{ if } -1 \leq n \leq 1$$
$$= 0, \text{ otherwise} \quad (18)$$

The radial basis function is given by

$$\phi(x) = \exp(-\frac{\|x-\mu\|^2}{\sigma^2}) \quad (19)$$

The procedure of ELM for single-layer feed forward networks is expressed as follows:

1) Choose arbitrary value for input weights w_i and biases of hidden neurons b_i.
2) Calculate hidden layer output matrix H.
3) Obtain the optimal $\hat{\beta}$ using, $\hat{\beta} = H^\dagger T$.

1. Training and Testing: 10 Fold Cross Validation

Among the variety of methods to divide the EEG signals, to reduce the bias of training and testing data, a 10 fold cross validation method is used [19]. This method is implemented during the training period to estimate the classification model that learns from the training data. Usually, the data set is divided into 10 subsets, and holdout approach is reiterated 10 times. Each time, one of the 10 subsets is utilized as the testing dataset and 9 other subsets are combined to form training data set. At last, the average error for 10 trials is calculated.

2. Mean Square Error

The Mean Square Error (MSE) of the actual target and the observed target are calculated as in [20]

$$\text{MSE} = (T_j - O_j)^2 (i, j = 1, \ldots, k) \tag{20}$$

The obtained mean square error for the ELM algorithm with the different activation function and hidden neurons are given in Table 3.1- Table 3.3. The results show the ELM algorithm followed by Singular Value Decomposition for dimensionality reduction obtains minimum mean square error for all activation functions with 40 hidden neurons than the ELM followed by PCA and ICA. On comparing the MSE's of ELM with PCA and ELM with ICA, the ELM with ICA obtains minimum MSE in average.

Table 3.1: Average MSE of ELM with SVD dimensionality reduction

Hidden Neurons	Sigmoid	Sine	Hardlim	Tribas	Radbas
10	0.000006245	0.000014285	0.001174300	0.000098351	0.000021112
15	0.000010841	0.000030959	0.000910873	0.000138436	0.000052146
20	0.000006245	0.000014285	0.000789431	0.000098351	0.000021112
30	0.000000013	0.000000064	0.000003563	0.000008423	0.000000754
40	0.000000011	0.000000059	0.000003544	0.000003573	0.000000690
Average	**0.000004090**	**0.000013014**	**0.000482228**	**0.000068568**	**0.000017515**

Table 3.2. Average MSE of ELM with PCA dimensionality reduction

Hidden Neurons	Sigmoid	Sine	Hardlim	Tribas	Radbas
10	0.00051079	0.00056202	0.00054566	0.00138317	0.00056814
15	0.00000325	0.00006049	0.00006614	0.00046955	0.00012318
20	0.00000278	0.00006036	0.00003816	0.00014699	0.00005675
30	0.00000105	0.00006204	0.00003924	0.00024198	0.00007908
40	0.00000069	0.00008035	0.00001274	0.00011087	0.00004436
Average	**0.00010371**	**0.00016505**	**0.00014039**	**0.00047051**	**0.00017430**

Table 3.3. Average MSE of ELM with ICA dimensionality reduction

Hidden Neurons	Sigmoid	Sine	Hardlim	Tribas	Radbas
10	0.00000334	0.00002583	0.00009398	0.00039314	0.00003203
15	0.00000200	0.00002669	0.00006878	0.00010367	0.00001580
20	0.00000106	0.00002833	0.00008045	0.00008001	0.00001122
30	0.00000095	0.00002344	0.00005691	0.00005112	0.00000833
40	0.00000084	0.00002167	0.00006234	0.00003650	0.00000562
Average	**0.00000164**	**0.00002519**	**0.00007249**	**0.00013289**	**0.00001460**

4. Continuous Genetic Algorithm as a Post Classifier for Epilepsy Risk Level Classification

Genetic algorithm is based on biological concept of generation of the population, a rapid growing area of Artificial intelligence. GA's are inspired by Darwin's theory about Evolution. According to the Darwin "Survival of the fittest". Genetic Algorithm is based on the biological concept of population generation.

A typical genetic algorithm requires:
i. A genetic representation of the solution domain,
ii. A fitness function to evaluate the solution domain.

Once the genetic representation and the fitness function are defined, a GA proceeds to initialize a population of solutions and then to improve it through repetitive application of the mutation, crossover, inversion and selection operators. GA has blossomed rapidly due to the easy availability of low cost but fast speed small computers. The complex and conflicting problems that required simultaneous solutions, which in past were considered deadlocked problems, can now be obtained with GA. However, the GA is not considered a mathematically guided algorithm. The optima obtained are evolved from generation to generation without stringent mathematical formulation such as the traditional gradient–type of optimizing procedure. In fact, GA is much different in that context. It is merely a stochastic, discrete event and a non linear process. The obtained optima are an end product containing the best elements of previous generations where the attributes of a stronger individual tend to be carried forward into the following generation. The rule of the game is "survival of the fittest will win" [21].

Most values in the real world are neither integers nor in encoded form. They appear as real values and the advancement in computing has enabled us in realizing these values to a high precision. Hence there is requirement of an algorithm, which can be used in manipulating these real valued data. In a similar sense, the output of the fuzzy system is also a real number in the '0 to 20' scale with a precision of 4 digits after decimal point. Instead of encoding the fuzzy output in the form of strings and converting them to bit representation, we can directly use the values in a continuous parameter genetic algorithm (CGA) to obtain more precise outputs. The advantages of using a CGA are:

a. Reduction in memory size occupied
b. Reduction in quantization error
c. Increase in precision
d. Speedup of computation

The iterations in a continuous parameter genetic algorithm follow the footsteps of its predecessor. The difference in the continuous parameter genetic algorithm occurs in the computation of the fitness function and the crossover and mutation operators. The following section elucidates as to how the operators in a continuous parameter genetic algorithm differ from a conventional binary genetic algorithm.

4.1 Functions and Operators

Fitness function:
The fitness function is calculated on a 0 to 1 scale. As the output risk level can have a maximum value of 20, dividing by 20 normalizes each gene and the average of all six genes is taken as the fitness of the chromosome. This fitness value of the chromosome is used in selecting the chromosome for the next level of iteration after sorting.

Crossover:
A single point crossover with a random generation of crossover point from the 2nd to the 5th position is followed. As the iteration involves real numbers, a different method for crossover is followed which is illustrated below using symbols.

$Parent_1 = [\ C_{11},\ C_{12},\ C_{13}.\ C_{14},\ C_{15},\ C_{16}]$

$Parent_2 = [\ C_{21},\ C_{22},\ C_{23}.\ C_{24},\ C_{25},\ C_{26}]$

$Offspring_1 = [\ C_{11},\ C_{12},\ C_{13}.\ C_{X4},\ C_{X5},\ C_{X6}]$

$Offspring_2 = [\ C_{21},\ C_{22},\ C_{23}.\ C_{Y4},\ C_{Y5},\ C_{Y6}]$

where, $C_{Xn} = \beta C_{1n} + (1-\beta) C_{2n}$

$C_{Yn} = \beta C_{2n} + (1-\beta) C_{1n}$

β = random number in [0, 1]

The level of crossover varies with the value of β. For $\beta=1$, there is no crossover and for $\beta=0$, there is a complete crossover. For fractional values of β, there is a corresponding change in the two strings. The crossover rate used in this project is 0.5.

Mutation:

A random point mutation is done by selecting one of the chromosomes and changing its gene to a value in [0, 20] which is the output risk level of the system. This mutation occurs only if a random number generated is less than the probability of mutation. The mutation rate if fixed as 0.1.

A simple genetic algorithm can be summed up in seven steps as follows [22]:Start with a randomly generated population of n chromosomes
2. Calculate fitness of each chromosome
3. Select a pair of parent chromosomes from the initial population
4. With a probability P_{cross} (the 'crossover probability' of the 'crossover rate'), perform crossover to produce two offspring
5. Mutate the two offspring with a probability P_{mut} (the mutation probability)
6. Replace the offspring in the population
7. Check for termination or go to step 2

Each iteration of the above steps is called a generation. The termination condition is usually a fixed number of generations typically anywhere from 50 to 500 or more. Under certain other circumstances, a check for global minimum is done after each generation and the algorithm is terminated as and when it is reached [23].

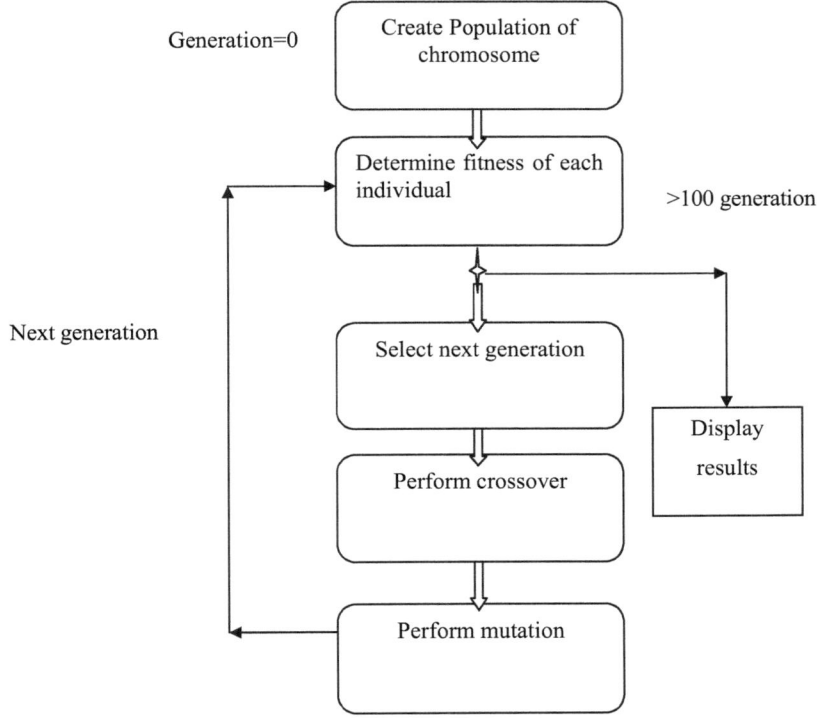

Figure 4.1: Flow Diagram for Continuous Genetic Algorithm

4.2 Mean Square Error

The mean Square Error is calculated by using the equation (3.10). The mean square error for Continuous Genetic algorithm with the SVD , PCA and ICA Dimensionality Reduction techniques are given in Table 4.1. We can see from the table that the GA with SVD dimensionality Reduction obtains less Mean Square Error compared to the PCA or ICA dimensionality reductions. Figure 4. 2 shows MSE for GA with various numbers of generations and it proves that as the number of generation increases to 128 the Mean Square Error decreases.

Table 4.1: Average MSE for GA with SVD, PCA and ICA Techniques

PATIENT	MSE		
	GA	GA (PCA)	GA (ICA)
1	0.0002971	0.0002560	0.0002914
2	0.0001200	0.0002761	0.0002529
3	0.0000834	0.0000874	0.0001775
4	0.0001236	0.0003118	0.0003583
5	0.0000609	0.0001774	0.0001548
6	0.0000930	0.0002940	0.0002782
7	0.0001673	0.0002105	0.0001353
8	0.0003906	0.0001539	0.0001903
9	0.0001588	0.0002277	0.0001165
10	0.0001421	0.0002662	0.0001823
11	0.0002086	0.0002415	0.0001622
12	0.0003258	0.0002047	0.0003976
13	0.0002517	0.0001176	0.0001982
14	0.0002922	0.0002632	0.0004151
15	0.0001645	0.0003941	0.0002781
16	0.0003010	0.0002831	0.0004725
17	0.0002924	0.0002943	0.0005944
18	0.0001310	0.0003503	0.0002022
19	0.0002517	0.0002514	0.0002980
20	0.0001200	0.0002761	0.0002529

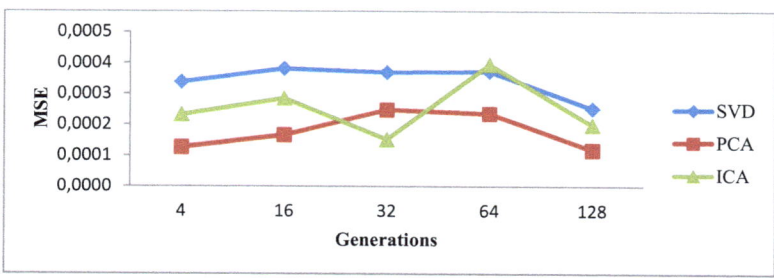

Figure 4.2: MSE for GA with various numbers of generations of patient 13

5. Results and Discussion

The tool that has been used in this paper is MATLAB 10. Input signal from twenty set of patients each patient is considered. There are four epoch for each patient, each epoch consisting of 16 channels. Thus there are around 64 channels for each patient under treatment. The total number of samples processed is 5, 12,200. The section below shows the results and their discussion.

5.1 Performance Index

The Performance Index is defined as in [24]

$$PI = \frac{PC - MC - FA}{PC} \times 100 \tag{5.1}$$

Where PC – Perfect Classification, MC – Missed Classification, FA – False Alarm

The perfect classification represents when the physician agrees with the epilepsy risk level. Missed classification represents a High level as Low level. False alarm represents a Low level as High level with respect to physician's diagnosis. The other performance measures such as sensitivity (S_e) and specificity (S_p) as in [25] are calculated using (5.2) and (5.3).

$$S_e = [PC/(PC+FA)]*100 \tag{5.2}$$

$$S_p = [PC/(PC+MC)]*100 \tag{5.3}$$

5.2 Quality Value

To reflect the overall quality of the classifier, the Quality Values are calculated using the three factors such as Classification rate, Classification delay, and False Alarm rate. The quality value Q_V is defined as [26],

$$Q_V = \frac{C}{(R_{fa} + 0.2)*(T_{dly}*P_{dct} + 6*P_{msd})} \qquad (5.4)$$

Where,

- C is the scaling constant,
- R_{fa} is the number of false alarm per set
- T_{dly} is the average delay of the on set classification in seconds
- P_{dct} is the percentage of perfect classification and
- P_{msd} is the percentage of perfect risk level missed.

A constant C is empirically set to 10 because this scale is the value of Q_V to an easy reading range the classifier with higher Q_V is the better one.

Table 5.1: Performance Analysis of ELM with SVD dimensionality reduction and various hidden neurons

Hidden Neurons	PC	MC	FA	Sensitivity	Specificity	PI	QV	Time Delay	Average Detection
10	96.456	2.2049	1.3312	98.5814	97.8733	96.33	22.74	2.061	98.22735
15	96.331	2.6626	0.9984	98.8308	97.498	96.20	22.82	2.086	98.1644
20	96.539	2.3298	1.1232	98.7057	97.8316	96.42	22.86	2.07	98.26865
30	97.206	2.3296	0.4576	99.5413	97.6648	97.13	23.45	2.083	98.60305
40	97.373	2.2048	0.416	99.583	97.7899	97.31	23.55	2.079	98.68645

The ELM with SVD dimensionality Reduction having 40 hidden neurons obtains high performance index, sensitivity, specificity and average detection [27]. The ELM with PCA dimensionality reduction obtains a high performance index of 92.93 with 15 hidden neurons. . The ELM with ICA dimensionality reduction obtains a high performance index of only 87.9 with 30 hidden neurons. The overall performance of the ELM is best with SVD as the technique for dimensionality reduction. Figure 5.1 shows the Quality Value for ELM with SVD, PCA and ICA with different hidden neurons. We can observe that the ELM with SVD obtains high quality value with 40 hidden neurons followed by ELM with PCA and ICA dimensionality Reductions.

Table 5.2: Performance Analysis of ELM with PCA dimensionality reduction and various hidden neurons

Hidden Neurons	PC	MC	FA	Sensitivity	Specificity	PI	QV	Time Delay	Average Detection
10	92.85	4.14	3.01	96.99	95.86	92.29941	20.64	2.1054	96.425
15	93.43	4.8515	1.75	98.25	95.18	92.9342	21.29	2.1596	96.715
20	91.44	6.55	2.01	97.99	93.45	90.63867	20.45	2.2218	95.72
30	90.35	7.87	1.78	98.22	92.13	89.31931	20.14	2.2792	95.175
40	90.16	7.12	2.72	97.28	92.88	89.08607	19.73	2.2304	95.08

Table 5.3: Performance Analysis of ELM with ICA dimensionality reduction and various hidden neurons

Hidden Neurons	PC	MC	FA	Sensitivity	Specificity	PI	QV	Time Delay	Average Detection
10	86.28	11.63	2.09	97.91	88.37	84.09828	18.68	2.4234	93.14
15	86.81	12.28	0.91	99.09	87.72	84.8059	19.34	2.473	93.405
20	86.5	12.93	0.57	99.43	87.07	84.39306	19.4	2.5058	93.25
30	87.6	12.1	0.3	99.7	87.9	85.84475	19.88	2.478	93.8
40	88.4	10.9	0.7	99.3	89.1	86.87783	19.95	2.422	94.2

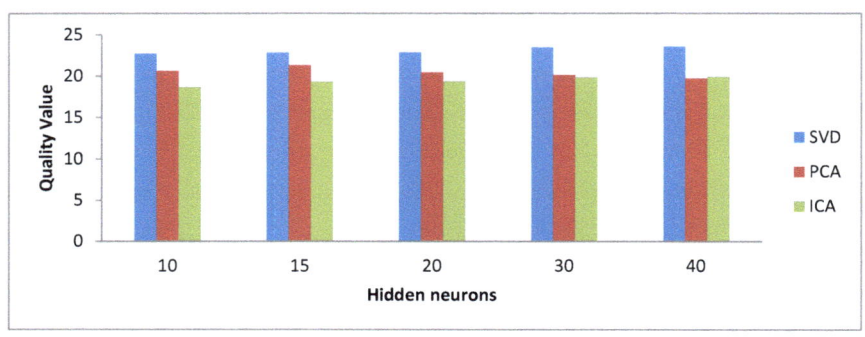

Figure 5.1: Quality Value for ELM with SVD, PCA and ICA with different hidden neurons

The Performance analysis of GA classifier using SVD, PCA and ICA are given in table 5.4 . The table shows that GA with SVD dimensionality Reduction outperforms PCA and ICA in PI, QV, Sensitivity and Specificity. Figure 5.2 shows the Quality Value for Continuous GA with SVD, PCA and ICA. We can see that GA with SVD dimensionality reductions has high quality value as 19.51 compared to the others. But this value is less than the Quality Value obtained by the ELM classifier.

Table 5.4: Performance Analysis of Continuous GA with dimensionality reductions

Techniques	PC	MC	FA	Sensitivity	Specificity	PI	QV	Time Delay	Average Detection
SVD	89.14	2.63	8.22	91.77	97.36	87.15	19.51	1.94	94.572
PCA	85.96	6.57	7.45	92.54	93.42	83.18	18.61	2.11	92.982
ICA	87.71	2.412	9.86	90.13	97.58	85.28	18.84	1.89	93.859

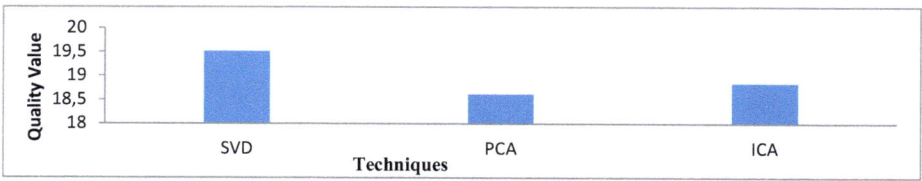

Figure 5.2: Quality Value for Continuous GA with SVD, PCA and ICA

The Sensitivity & Specificity graph of various activation functions for ELM with SVD, PCA ICA dimensionality reduction methods are given in Figure 5.3-5.5. The Sensitivity and Specificity graph for GA with SVD, PCA ICA dimensionality reduction methods with 40 hidden neurons are given in Figure 5.6.

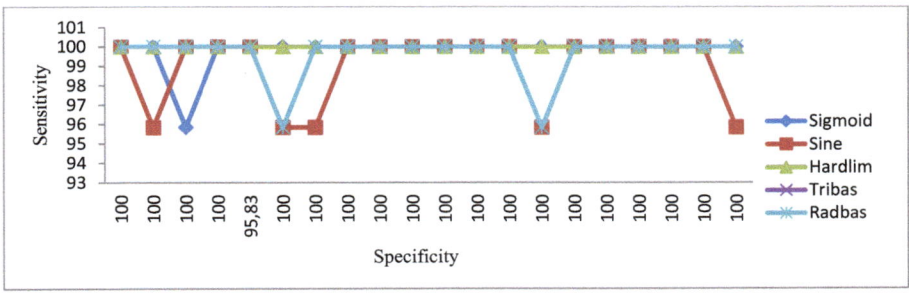

Figure 5.3: Sensitivity and Specificity measures of ELM using SVD Dimensionality Reduction (40 Hidden neurons)

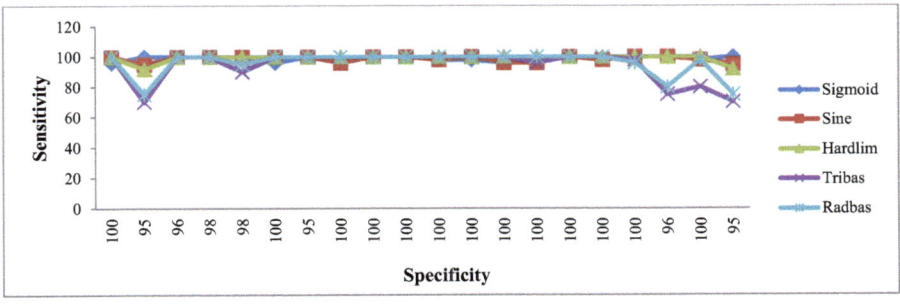

Figure 5.4: Sensitivity and Specificity measures of ELM using PCA Dimensionality Reduction (40 Hidden neurons)

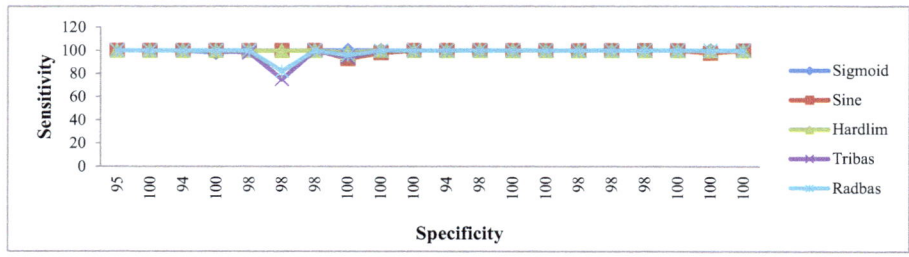

Figure 5.5: Sensitivity and Specificity measures of ELM using ICA Dimensionality Reduction (40 Hidden neurons)

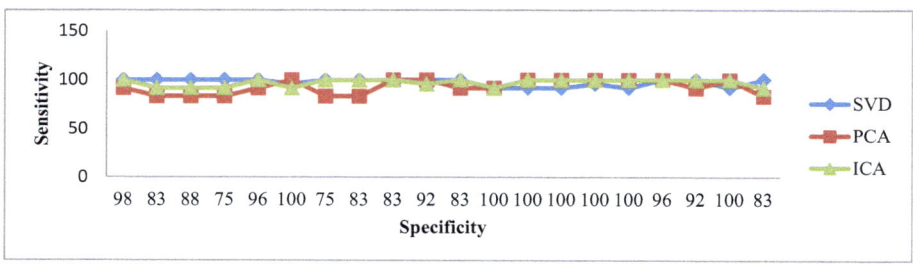

Figure 5.6: Sensitivity and Specificity measures of Continuous GA using SVD, PCA and ICA

The Time Delay and Quality Value for measures for the ELM classifier with SVD, PCA and ICA dimensionality reduction having 49 hidden neurons are given in Figure 5.7-5.9.

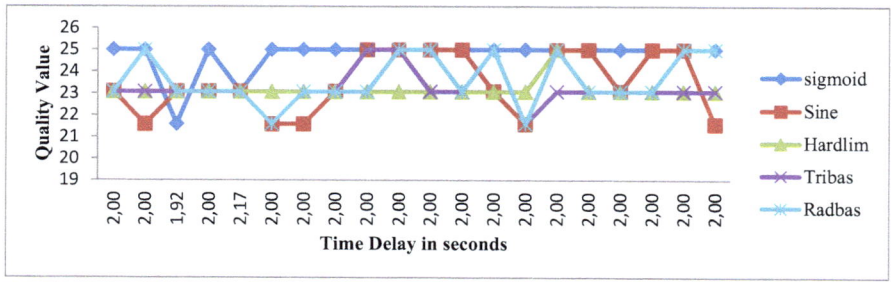

Figure 5.7: Time Delay and Quality Value Measures of ELM Classifier with SVD dimensionality reduction of 40 hidden neurons

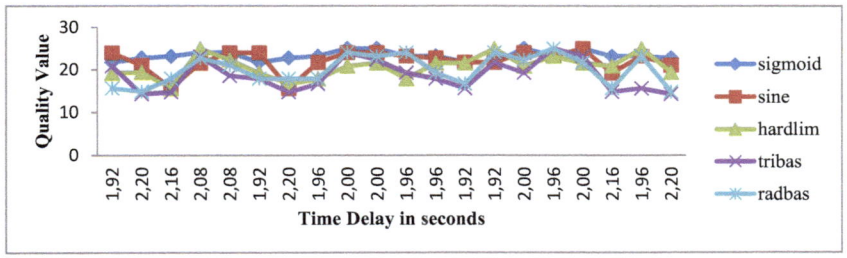

Figure 5.8: Time Delay and Quality Value Measures of ELM Classifier with PCA dimensionality reduction of 40 hidden neurons

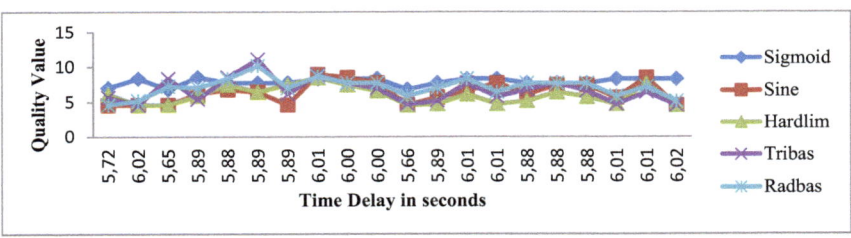

Figure 5.9: Time Delay and Quality Value Measures of ELM Classifier with ICA dimensionality reduction of 40 hidden neurons

We can observe that Time delay and the Quality Value measures of the SVD dimensionality reduces ELM classifier performs best. This observation is suitable for the Continuous Genetic Algorithm classifier also.

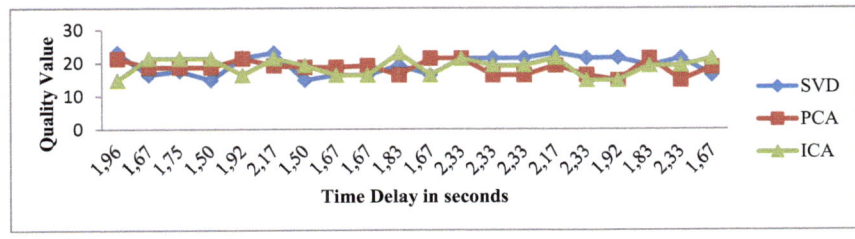

Figure 5.10: Time Delay and Quality Value Measures of Continuous GA Classifier with SVD, PCA and ICA dimensionality reduction

Table 5.5 compares the average performance of the ELM and Continuous GA as the post classifier with SVD, PCA and ICA as Dimensionality Reduction techniques. The high Performance Index and a Quality Value of 96.68 and 23.08 are obtained with ELM classifier with SVD dimensionality reduction. Continuous Genetic Algorithm with SVD dimensionality reduction obtains a high Performance Index of 87.15 and Quality value of 19.51 which is considerably less than the ELM classifier. Thus the results prove that ELM performance best than the Continuous GA in terms of Performance Index, Quality Value, Sensitivity and Average Detection.

Table 5.5: Performance Analysis of ELM and Continuous GA

Parameters	ELM			GA		
	SVD	PCA	ICA	SVD	PCA	ICA
Perfect Classification	96.78	91.65	87.12	89.14	85.96	87.71
Missed Classification	2.35	6.11	11.97	2.63	6.57	2.41
False Alarm rate/set	0.87	2.25	0.91	8.22	7.45	9.86
Sensitivity	99.05	97.75	99.09	91.77	92.54	90.13
Specificity	97.73	93.90	88.03	97.36	93.42	97.58
Performance Index %	96.68	90.86	85.20	87.15	83.18	85.28
Quality Value	23.08	20.45	19.45	19.51	18.61	18.84
Time Delay in seconds	2.08	2.20	2.46	1.94	2.11	1.89
Average Detection	98.39	95.82	93.56	94.57	92.98	93.86

6. Conclusion

This project analyses the performance of the Extreme Learning Machine and Continuous Genetic Algorithm in optimizing the epilepsy risk level of epileptic patients from EEG signals. The dimensionality reduction is done by Singular value Decomposition, Principal Component Analysis and Principal Component Analysis. The results prove that ELM performance best than the Continuous GA in terms of Performance Index, Quality Value, Sensitivity and Average Detection. From this method, the risk levels of the patients are identified and proper medication can be given to them.

References

[1] R.Harikumar, Dr.(Mrs.). R.Sukanesh, P.A. Bharathi, Genetic Algorithm Optimization of Fuzzy outputs for Classification of Epilepsy Risk Levels from EEG signals, Journal of Interdisciplinary panels I.E. (India), vol.86, no.1, May 2005, pp1-10.

[2] Seunghan Park , *"TDAT Domain Analysis Tool for EEG Analysis"*, IEEE Transactions on Biomedical Engineering, 37(8) pp 803-811, August 1990.

[3] R.Neelaveni and G.Gurusamy, "EEG Signal Analysis Methods – A Review*",* Proceedings of National System Conference, pp. 355-361, 1998.

[4] Kirk Baker, "Singular Value Decomposition Tutorial", March 29, 2005.

[5] Michael E.Wall,Andreas Rechtsteiner,Luis M.Rocha, "Singular value decomposition and principal component analysis: in a practical approach to microarray data analysis" Kluwer: Norwell, MA, 2003. pp. 91-109.

[6] Lindsay I Smith, "A tutorial on Principal Components Analysis", February 26, 2002.

[7] Michael E_ Tipping, Christopher M_ Bishop, "Probabilistic principal component analysis" Neural Computing Research Group, September 4 , 1994.

[8] Penny W., Everson R., Roberts S. "Hidden Markov Independent Components Analysis", 2000. *URL citeseer.ist.psu.edu*/397426.html.

[9] Lee T-W. Girolami M., Bell A.J. and Sejnowski T.J,"A unifying Information-theoretic framework for Independent Component Analysis",*International Journal on Mathematical and Computer Modeling*, in press.

[10] Aapo Hyvärinen and Erkki Oja, "Independent Component Analysis: Algorithms and Applications", *Neural Networks, 13(4-5):411-430, 2000.*

[11] G.B. Huang ,Q.Y. Zhu and C.K.Siew, "Extreme learning machine : A new learning scheme of feedforward neural networks" in proceedings of international joint conference on neural networks.(IJCNN 20004)(budapest, Hungary),25-29 July, 2004.

[12] G.B. Huang and C.K.Siew, "Extreme learning machine: RBF network case", in proceedings of eighth international conference on control automation, robotics and vision(ICARCV 2004), (Kunming, China) , 6-9 December 2004.

[13] G.B. Huang and C.K.Siew, "Extreme learning machine with randomly assigned RBK kernels", international journal of information technology, Vol.11, no.1,2005.

[14] Wang D. and Huang, G.B. (2005), "Protein sequence classification using extreme learning machine". Proceedings of the 2005 IEEE International Joint Conference on Neural Networks, Montreal, 3, 31 July-4 August 2005, 53-59.

[15] R. Zhang, R., Huang, G.B., Sundararajan, N. and Saratchandran.P, "Multicategory classifcation using an extreme learning machine for microarray gene expression cancer diagnosis" IEEE Transactions on Computational Biology and Bioinformatics, 2007 .4(3), 485-495.

[16] Liang, N.Y., Saratchandran, P., Huang, G.B. and Sundararajan, N. "Classification of mental tasks from EEG signals using extreme learning machine". International Journal of Neural Systems 2006, 16(1), 29-38.

[17] Guang-Bin Huang and Haroon A. Babri, "Upper Bounds on the Number of Hidden Neurons in Feedforward Networks with Arbitrary Bounded Nonlinear Activation Functions" , IEEE Transactions On Neural Networks, Vol. 9, No. 1, January 1998.

[18] Guang-Bin Huang, Qin-Yu Zhu, K. Z. Mao, K. Z. Mao, P. Saratchandran, N. Sundararajan, "Can Threshold Networks be Trained Directly?", IEEE Transactions On Circuits And Systems—II: Express Briefs, Vol. 53, No. 3, March 2006.

[19] Y.Song ,P.Lio, "A new approach for epileptic seizure detection: sample entropy based feature extraction and extreme learning machine", J. Biomedical Science and Engineering , 2010, 3, 556-567.

[20] Ye Yuan, Detection of Epileptic Seizure Based on EEG Signals,3rd International Congress on Image and Signal Processing (CISP2010), IEEE: pp 4209 – 4211.

[21] D.E.Goldberg, "Genetic Algorithms in Search, Optimization and Machine Learning", Reading, MA: Addison-Wesley, 1989.

[22] Marco Russo, "FuGeNeSys – A Fuzzy Genetic Neural System for Fuzzy Modeling", IEEE Transactions on Fuzzy Systems, 6(3) pp 373 – 387, August 1998

[23] Melanie Mitchell, "An Introduction to Genetic Algorithms", A Bradford Book MIT Press, 1997.

[24] Dingle, Alison A. Jones, Richard D. Carroll, Grant J,Fright, Richard Richard , A Multistage System to Detect Epileptic form Activity in the EEG, IEEE Transactions on Biomedical Engineering, Vol 40,No.12, pp 1260-1268, December 1993.

[25] R.Harikumar, Dr.(Mrs.). R.Sukanesh, P.A. Bharathi, Genetic Algorithm Optimization of Fuzzy outputs for Classification of Epilepsy Risk Levels from EEG signals, Journal of Interdisciplinary panels I.E. (India), vol.86, no.1, May 2005, pp1-10.

[26] R.Harikumar and B.Sabarish Narayanan, "Fuzzy Techniques for Classification of Epilepsy risk level from EEG Signals", Proceedings of IEEE Tencon – 2003, Bangalore, India, pp 209-213, 14-17 October 2003.

[27] M.Balasubramani, Dr.R.Harikumar, Dr.C.Ganeshbabu and G.A.Nivhedhitha, "Performance Analysis of Extreme Learning Machine for Robust Classification of Epilepsy Detection from EEG Signals", INFORMATION- An International Interdisciplinary Journal ,Vol.17, No.4, April, 2014 pp.1313-1324. Annexure I, 3635.